ANIMAL SUPERPOWERS!

WRITTEN BY
AMY CHERRIX

ILLUSTRATED BY
FRANN PRESTON-GANNON

Beach Lane Books • New York London Toronto Sydney New Delhi

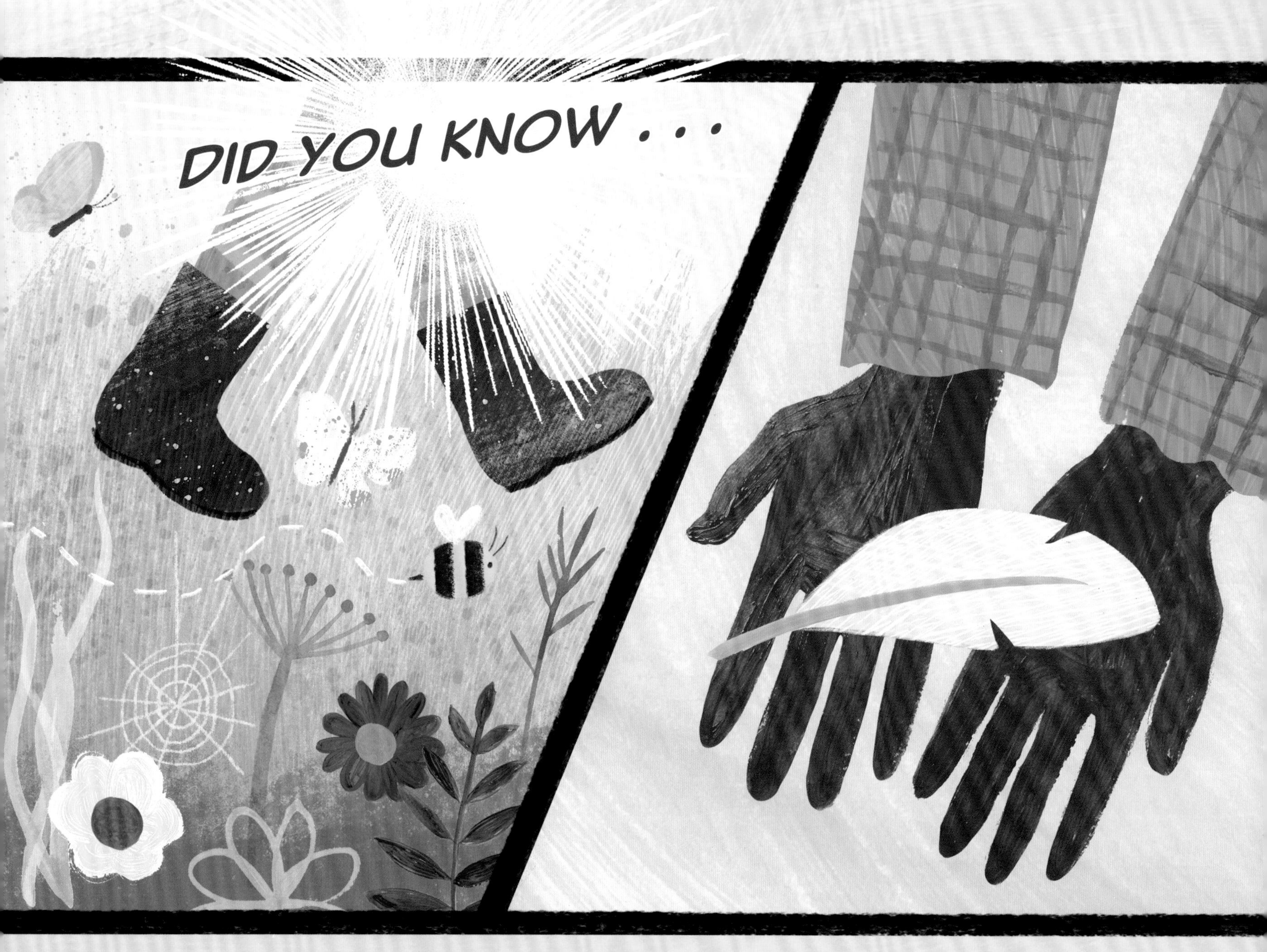

that many of Earth's animals have superpowers? These behaviors and physical traits are called adaptations, and animals use them in amazing ways. Some are expert escape artists. Others subdue their prey with electricity—or bubbles!

Come meet these curious creatures and explore how they use their natural superpowers to survive.

What if you could live forever? This pea-sized jellyfish, called *Turritopsis*, may know the answer. It appears to possess the superpower of . . .

IMMORTALITY.
When injured, this jellyfish can reset its life cycle. First, it sinks to the seafloor, where the medusa, or "bell," folds into itself and becomes a ball of cells, or "cyst." Gradually, something miraculous happens. Rootlike stems called "stolons" emerge from the shell.

As the stolons lengthen, tiny buds called "polyps" appear. This is a polyp colony. The buds on the polyp colony bloom into brand-new medusa bells. Where there was once only one jellyfish, now there are many, each with a new life to live.

In Northern Italy, a young alpine ibex forages for the salt and other minerals that her body needs. The search for nutrients will take this kid to death-defying heights! Her superpower is . . .

DEFYING GRAVITY.

The kid follows in her mother's sure-footed steps, steadily moving upward and sideways across the brickwork face of the Cingino Dam. The rubbery split hooves of the ibex grip the surface like suction cups, making the animal less likely to fall. Up, up, up, the little kid goes, learning to climb, learning to survive.

Stand back! A hungry snapping shrimp is about to attack an unsuspecting krill . . .

SNAP
SNAP

WITH BUBBLES!

When prey approaches, the shrimp opens its giant snapper claw. Quickly, it slams shut, and a bubble bursts out at sixty miles per hour. The bubble collapses with a piercing “POP!”—a sound measured at an ear-splitting 220 decibels. (A thunderclap is only 120 decibels.) In an instant, the pressure from the blast kills the krill. It was no match for the snapping shrimp’s superloud bubble bomb.

A deer mouse skitters through the tall grass in search of food. The soft rustling sound captures the attention of a barn owl. The hungry hunter is a nearly inescapable predator, because it possesses the power of . . .

The barn owl sports a ruff of stiff feathers around its face. Like a satellite dish, they direct sound to the ear openings on either side of the owl's head.

Locked on to the mouse's location, the hunter makes its move. The mouse is none the wiser as the barn owl swiftly closes the distance between them. Legs extended, talons grasping, the great bird strikes! The attack is quick . . . and quiet.

Beneath the turbid waters of the Amazon River swims a fish teeming with the shocking superpower of . . .

The eel has poor vision and must rely on its nervous system to detect the slightest movements of prey. When a fish swims by, thousands of special cells in the eel's body react. Like a built-in battery pack, these electrocytes generate six hundred volts of electricity. (That's five times stronger than the 120-volt wall socket in your home.) In a flash, the eel zaps the fish, swallows it whole, and swims away.

If you could view the pencil-dot-size tardigrade through a microscope, you would see an animal that resembles a chubby bear with eight legs. But this funnel-faced charmer is tougher than it looks. The tardigrade is able to inhabit every continent on Earth, from the coldest to the hottest, because it is . . .

A SUPER SLEEPER.
An extreme form of self-protection called "cryptobiosis" helps the tardigrade survive when its environment changes. First, the legs retract and it curls into a tight ball, squeezing any excess water from its segmented body. All that remains is a dry husk, or "tun."

A tardigrade can live in its tun state for as long as ten years, making it one of the most durable forms of life on Earth. Scientists think they have been around for more than five hundred million years. (In case you're counting, that's older than the dinosaurs!)

A crab spider is about to create something wondrous.
It’s not a web or a cocoon. It’s a balloon! Why?
This crafty arachnid’s superpower is . . .

FLYING WITHOUT WINGS!
Using their "balloons," these spiders have been known to reach altitudes higher than two miles! They begin by releasing an anchor silk. Then the spider raises its front legs into the air. Each leg is covered in special hairs that feel for the Earth's electrical field and air currents, factors that influence spider flight.

Several more silk fibers are released, fanning out behind the spider's body. When it's time for takeoff, the anchor line is released, and it's bon voyage to the ballooning eight-legged aviator!

In the wetlands of Florida's Everglades, an adult opossum snacks on insects at the base of a tree. *Ssssss . . . rattle, rattle . . . sssss. . . .* A five-foot-long eastern diamondback rattlesnake slithers close. Is this the end of the line for the pink-nosed marsupial? Not this time! The opossum has a snake-fighting superpower that . . .

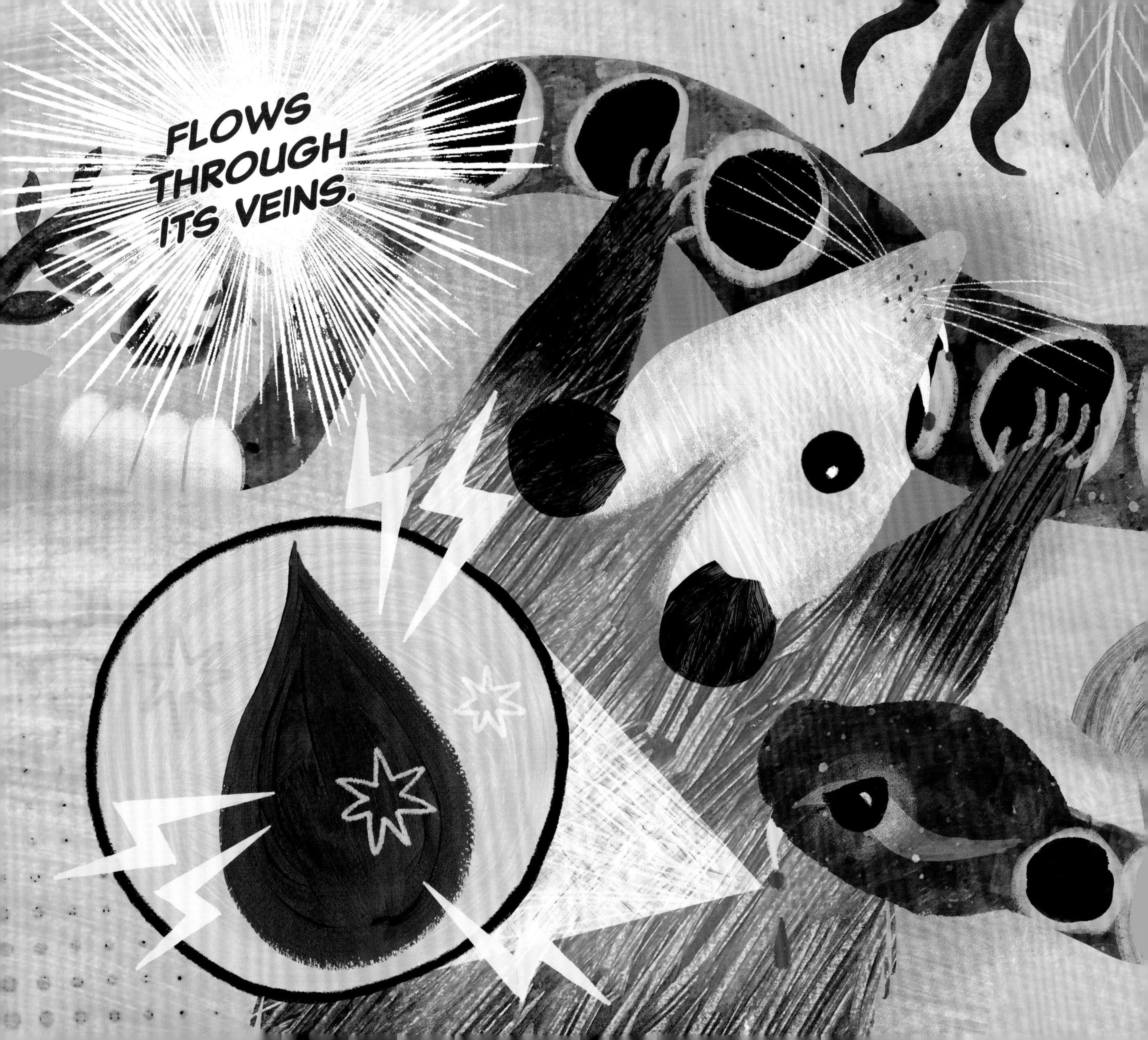
FLOWS THROUGH ITS VEINS.

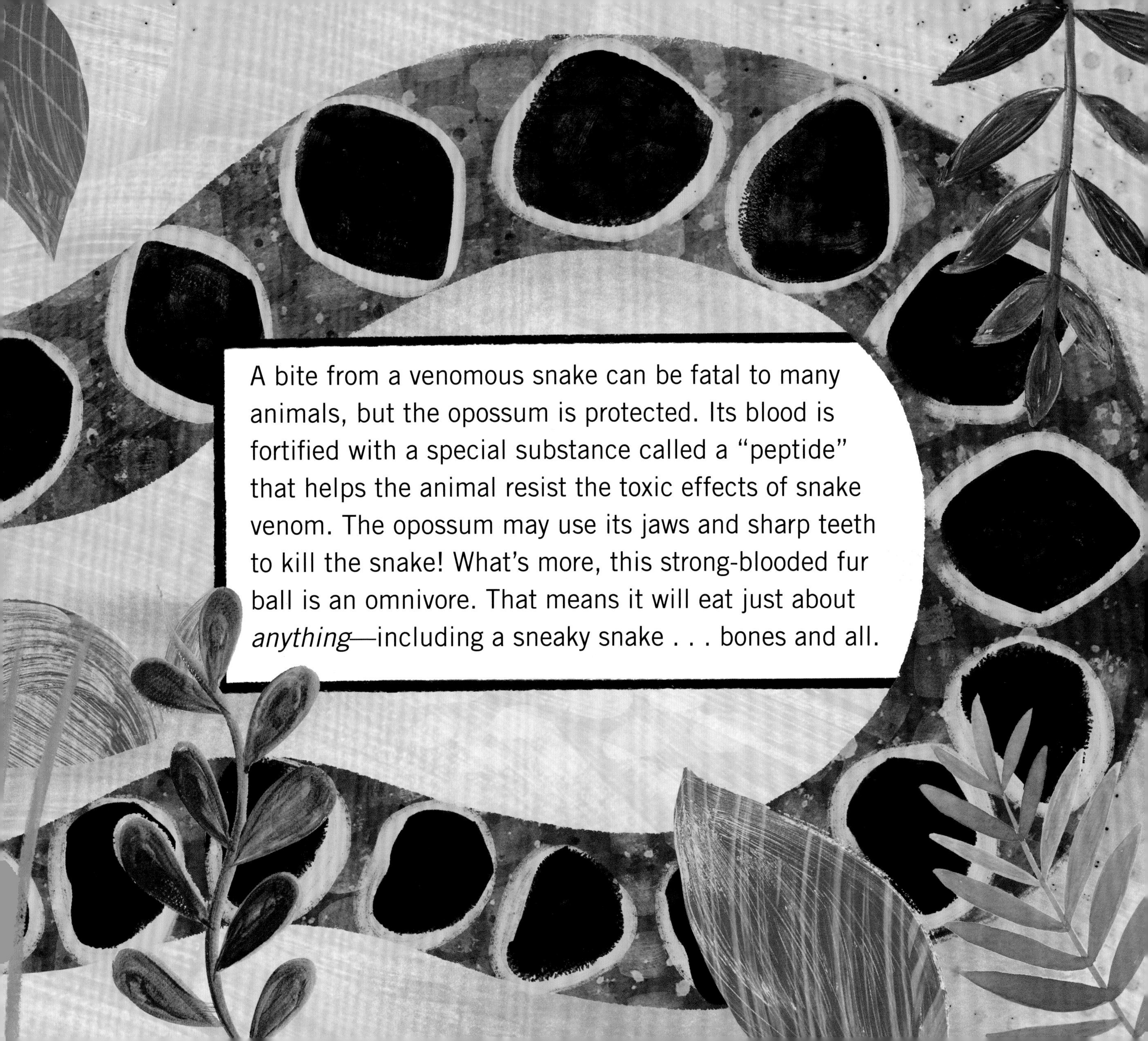

A bite from a venomous snake can be fatal to many animals, but the opossum is protected. Its blood is fortified with a special substance called a "peptide" that helps the animal resist the toxic effects of snake venom. The opossum may use its jaws and sharp teeth to kill the snake! What's more, this strong-blooded fur ball is an omnivore. That means it will eat just about *anything*—including a sneaky snake . . . bones and all.

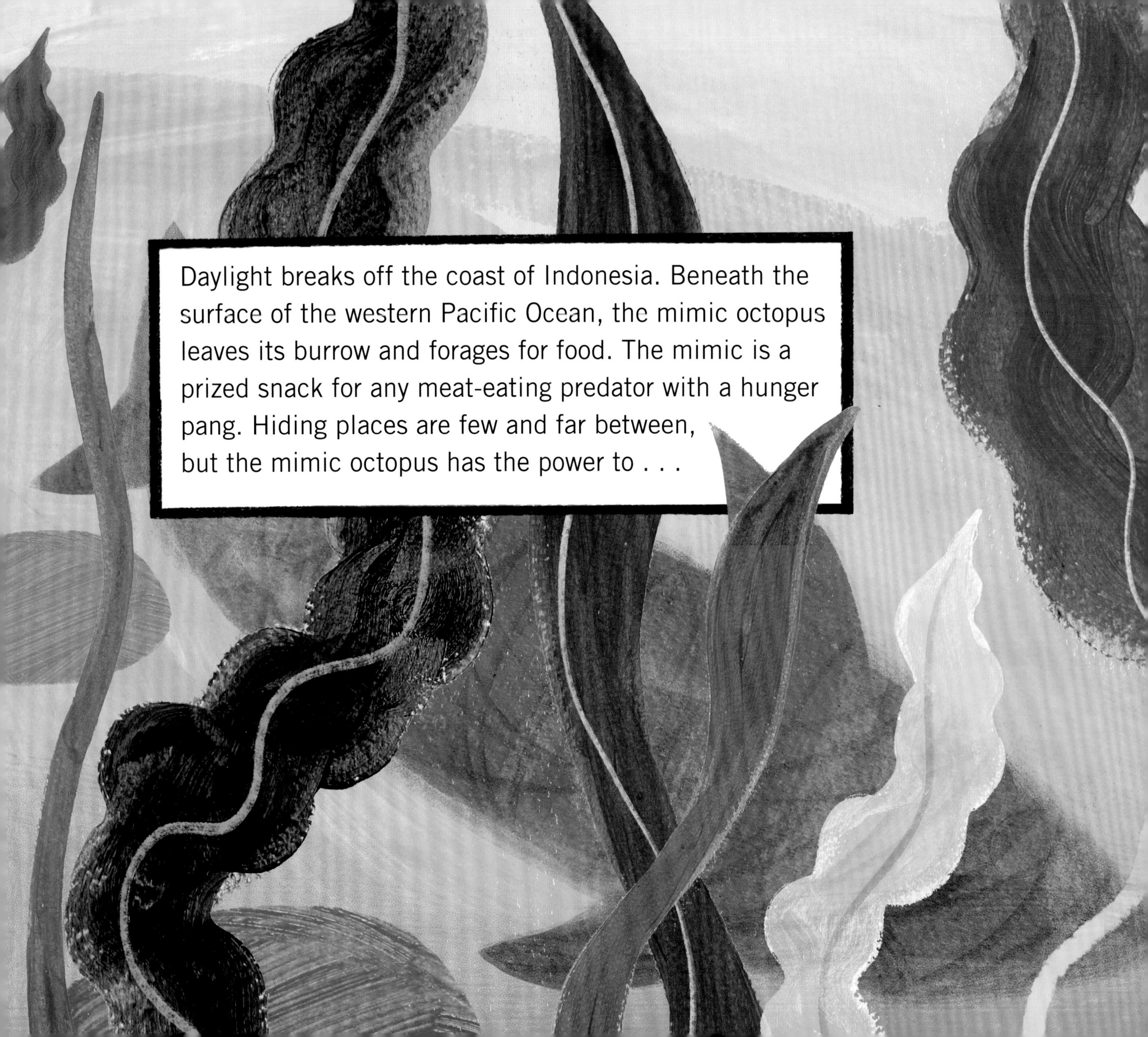
Daylight breaks off the coast of Indonesia. Beneath the surface of the western Pacific Ocean, the mimic octopus leaves its burrow and forages for food. The mimic is a prized snack for any meat-eating predator with a hunger pang. Hiding places are few and far between, but the mimic octopus has the power to . . .

HIDE IN PLAIN SIGHT.
Octopus flesh contains "chromatophores." These specialized cells transform the octopus's body in dynamic ways—altering its color or texture to match its surroundings.
But the mimic octopus may take its shape-shifting superpower an astonishing step further. When pursued by a predator, it appears to impersonate other—more threatening—animals in its habitat!

Suddenly, what looked and moved like an octopus now looks and moves like something else entirely. Something venomous. Something dangerous! A lethal lionfish? A deadly banded sea snake? Outwitted by the octopus's dazzling display, a hungry predator may give up the chase. And the mimic—the ocean's most talented master of disguise—survives.

If you have a dog (or have been smelled by one), congratulations! You've met an animal with a superpower. Dogs are . . .

Their noses are superb smelling machines that contain three hundred million scent detectors called "olfactory receptors." (Humans only have about six million.) When conditions are just right, dogs have been known to smell a person from as far away as twelve miles! Search and rescue dogs locate missing and injured persons.

Airport detection dogs sniff out dangerous substances and materials. Conservation dogs hunt for endangered plants and animals and root out invasive species. While some of the animals in this book could be hard to find or difficult to see, a dog's superpower might be right under *your* nose.

All around the world, creatures large and small—on the land, in the air, or in the water—use their natural superpowers to survive, protecting themselves and preserving our amazing animal kingdom.

The Earth is threatened by human activities that we have the power to change. Plastic waste pollutes the oceans, and the planet's climate is changing due in part to the burning of fossil fuels used to power our vehicles. As a result, weather patterns have been disrupted. The world's oceans are warming. Glaciers are melting. And sea levels are on the rise. But we human beings can use our superpower of ACTION to make a difference. Can you find ways to improve the lives of animals in your habitat?

Here's how all the animals in this book face impacts from human activities now, or could in the future:

The super-surviving ***TURRITOPSIS,*** also known as the ***"IMMORTAL JELLYFISH,"*** may be an invasive species. It thrives in warm water, and as ocean temperatures increase, this jellyfish is expanding to other ocean habitats. The *Turritopsis* "hitches a ride" aboard cargo ships that crisscross the world, pumping seawater—and the tiny jellyfish—in and out of their ballast tanks along the way.

ALPINE IBEXES thrive in cooler climates. Global warming could make it more difficult for these animals to regulate their body temperature.

Scientists have observed that ***SNAPPING SHRIMP*** are louder in warmer water. As the oceans heat up due to climate change, these noisy crustaceans could raise the volume to levels that may interfere with the food-finding abilities of other sea life.

Climate change could make it more difficult for ***BARN OWLS*** to survive temperature extremes and may also impact the availability of prey species they rely on for food.

The ***ELECTRIC EEL***, or ***KNIFEFISH***, is not yet a threatened species. However, the Amazon rainforest that surrounds its aquatic habitat is threatened by fire and deforestation.

Even the nearly indestructible ***TARDIGRADE*** could be threatened by climate change. Researchers found that some species of tardigrades are at their most vulnerable when exposed to high temperatures over an extended period of time.

CRAB SPIDERS can be particularly sensitive to shifts in their environment. When temperatures and patterns of precipitation suddenly change, spider populations may suffer. The use of chemical pesticides by humans also threatens spider populations.

As the climate warms in once cooler geographic areas, the ***OPOSSUM*** population is expanding into new habitats where it had previously been too cold for them to survive. It remains to be seen whether their arrival in these new habitats will help or disrupt the ecosystems.

The ***MIMIC OCTOPUS*** is not currently listed as a threatened species, but knowledge about this octopus species remains limited. If we protect the world's ocean habitats, marine animals like the mimic octopus will have a better chance of survival.

Domesticated ***DOGS*** could face increased exposure to severe infections like Rocky Mountain spotted fever, heartworms, and Lyme disease. As the climate warms, disease-carrying insects, like mosquitoes and ticks, may expand in number and to new geographical areas in response to temperature and moisture.

SELECTED SOURCES

Immortal Jellyfish: Maria Pia Miglietta, PhD, Associate Professor, Department of Marine Biology, Texas A&M University at Galveston, email correspondence with the author, July 13, 2023, and August 18, 2023.

Alpine Ibex: Biancardi, Carlo Massimo, and Alberto Enrico Minetti. "Gradient limits and safety factor of Alpine ibex (*Capra ibex*) locomotion." *Hystrix, the Italian Journal of Mammalogy* 28, no. 1 (2017): 56–60. https://doi.org/10.4404/hystrix-28.1-11504.

Snapping Shrimp: Koukouvinis, Phoevos, Christoph Bruecker, and Manolis Gavaises. "Unveiling the physical mechanism behind pistol shrimp cavitation." *Scientific Reports* 7, no. 13994 (2017). https://doi.org/10.1038/s41598-017-14312-0.

Barn Owl: Roulin, Alexander. *Barn Owls: Evolution and Ecology with Grass Owls, Masked Owls and Sooty Owls.* Cambridge: Cambridge University Press, 2020.

Electric Eel (Knifefish): Allison Waltz-Hill, Senior Aquarist, New England Aquarium, telephone interview with the author, December 13, 2021.

Tardigrade: Miller, William R. "Tardigrades: Bears of the Moss." *Kansas School Naturalist* 43, no. 3 (1997). https://sites.google.com/g.emporia.edu/ksn/ksn-home/vol-43-no-3-tardigrades-bears-of-the-moss.

Crab Spider: Leadmon, Lauren, and Daniela Santamarina, "How Spiders Use Electricity to Fly | Decoder." National Geographic. Posted on April 25, 2019. YouTube video, 3:30. https://www.youtube.com/watch?v=Ja4oMFOoK50.

Opossum: Jennifer Knight, Education Director and Senior Naturalist, Balsam Mountain Trust, telephone interview with the author, January 20, 2022.

Mimic Octopus: Brett Grasse, Manager, Cephalopod Operations, Marine Biological Laboratory, telephone interview with the author, December 6, 2021.

Dog: Tyson, Peter. "Dogs' Dazzling Sense of Smell." PBS Nova. October 4, 2012. https://www.pbs.org/wgbh/nova/article/dogs-sense-of-smell/.

For a complete list of sources, visit amycherrix.com.

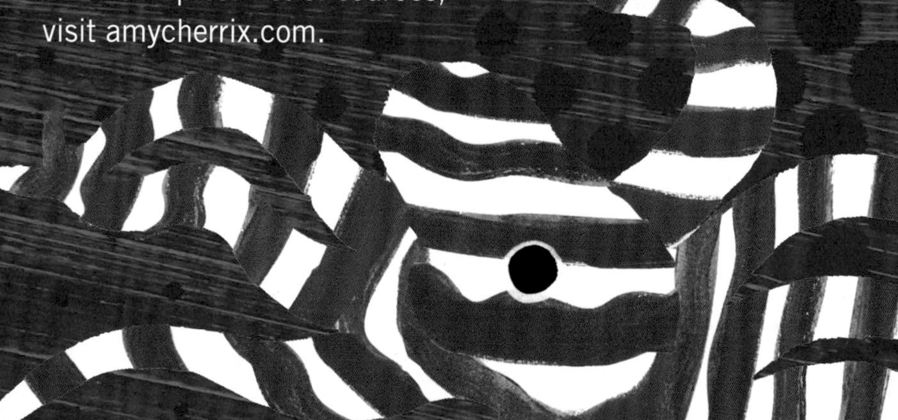

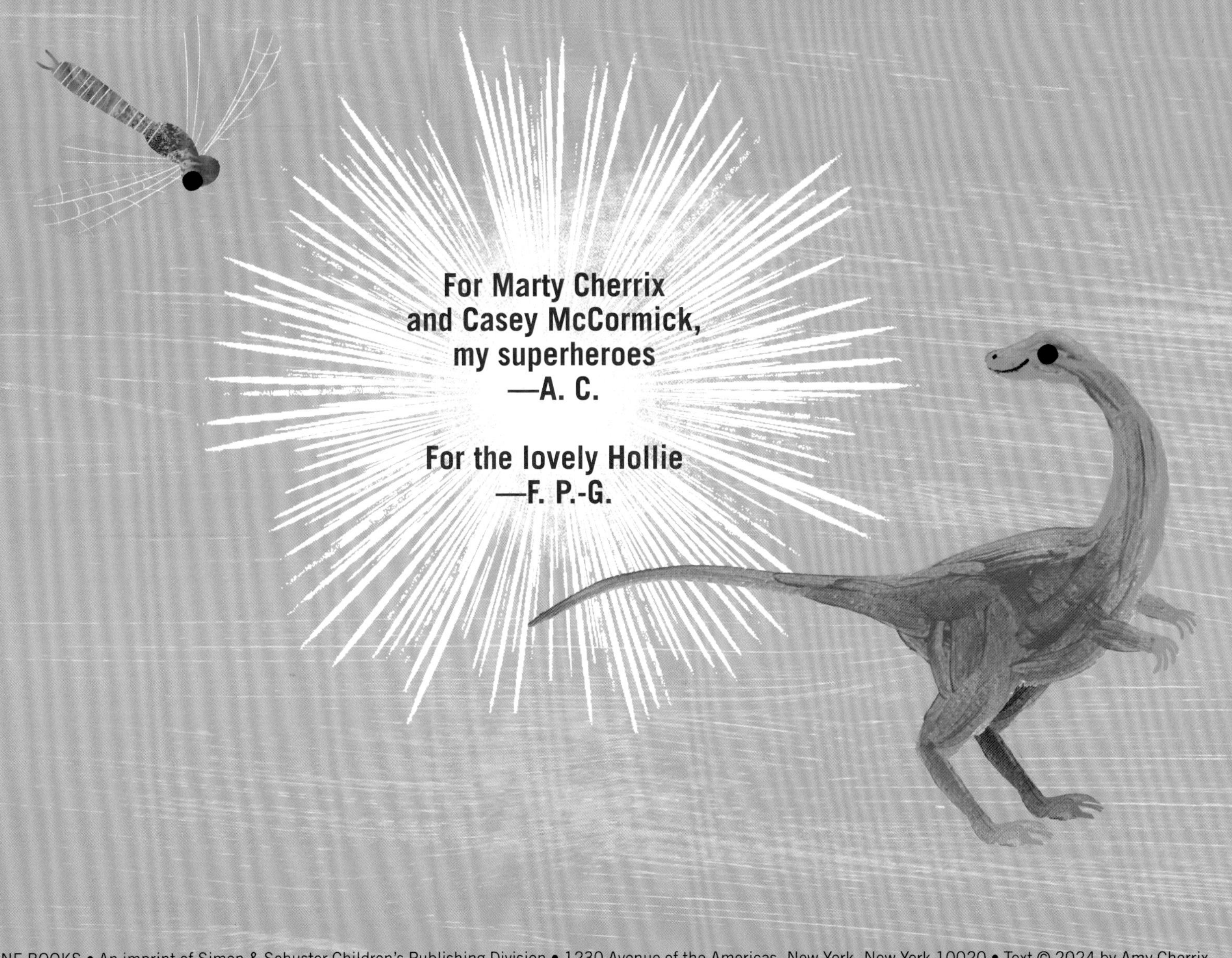

For Marty Cherrix
and Casey McCormick,
my superheroes
—A. C.

For the lovely Hollie
—F. P.-G.

BEACH LANE BOOKS • An imprint of Simon & Schuster Children's Publishing Division • 1230 Avenue of the Americas, New York, New York 10020 • • Simon & Schuster: Celebrating 100 Years of Publishing in 2024 • For information about special discounts for bulk purchases, please contact Simon & Schuster Special Sales at 1-866-506-1949 or business@simonandschuster.com. • The Simon & Schuster Speakers Bureau can bring authors to your live event. For more information or to book an event, contact the Simon & Schuster Speakers Bureau at 1-866-248-3049 or visit our website at www.simonspeakers.com. • The text for this book was set in Trade Gothic. • The illustrations for this book were rendered using a combination of acrylic paint and inks, which were then manipulated and collaged digitally using Photoshop. • Manufactured in China • 0224 SCP • First Edition • 10 9 8 7 6 5 4 3 2 1 • Library of Congress Cataloging-in-Publication Data • Names: Cherrix, Amy E., author. | Preston-Gannon, Frann, illustrator. • Title: Animal superpowers / Amy Cherrix ; illustrated by Frann Preston-Gannon. • Description: New York : Beach Lane Books, 2023. | Series: Amazing animals series | Includes bibliographical references. | Audience: Ages 0–8 | Audience: Grades 2–3 | Summary: "Did you know that many of Earth's animals have superpowers? These behaviors and physical traits are called adaptations, and animals use them in amazing ways. Some are expert escape artists. Other subdue their prey with electricity—or bubbles! Come meet these curious creatures and explore how they use their natural superpowers to survive"— Provided by publisher. • Identifiers: LCCN 2022008077 (print) | LCCN 2022008078 (ebook) | ISBN 9781534456273 (hardcover) | ISBN 9781534456280 (ebook) • Subjects: LCSH: Animals—Adaptation—Juvenile literature. • Classification: LCC QH546 .C44 2023 (print) | LCC QH546 (ebook) | DDC 591.4—dc23/eng/20220528 • LC record available at https://lccn.loc.gov/2022008077 • LC ebook record available at https://lccn.loc.gov/2022008078